Bibliographic information published by the German National Library:

The German National Library lists this publication in the National Bibliography; detailed bibliographic data are available on the Internet at http://dnb.dnb.de .

Imprint:

Copyright © 2015 GRIN Verlag, Open Publishing GmbH
Print and binding: Books on Demand GmbH, Norderstedt Germany
ISBN: 9783668329492

This book at GRIN:

http://www.grin.com/en/e-book/342816/specific-legal-issues-affecting-the-imple-mentation-of-computer-based-information

Daniel Kinyanjui

Specific legal issues affecting the implementation of computer-based information systems in the developing countries. A critical review of literature

GRIN Publishing

Article Tile

Specific legal issues affecting the implementation of computer-based information systems in the developing countries: a critical review of literature

Date: November 2015

About the Author

Daniel Kinyanjui (Bsc-Information Science-Moi University-Eldoret, Kenya and Currently Studying Masters of Library and Information Science at Kenyatta University-Kenya)

Daniel is also the editor of institutional in-house publications at Oshwal College-Nairobi and the College Head of Library Services.

Table of Contents

1.0 Abstract

Implementation of computer-based information systems (CBIS) has become increasingly important due to the growing reliance on new technologies by organisations in their operations and service delivery. There has also been realisation by individuals and businesses of the inevitability of delivering value-added services without computerized systems. Despite this awareness and the apparent growing readiness to deploy computer-based systems, there still exists a myriad of perennial challenges inhibiting the implementation of these systems. Using insights from previous studies, the paper sums up the legal issues that have been acknowledged by various scholars as major obstacles to the implementation of CBIS in developing countries.

Keywords and phrases: computer-based information systems, computers, developing countries,

Method. A review of English language literature published after 1998 to the present day relating to the legal challenges encountered in the implementation of computer- based information systems in developing countries

2.0 INTRODUCTION

According to Jessup and Valacich (2003), computer based information systems are "a combination of hardware, software, and telecommunications networks that people build and use to collect, create, and distribute useful data, typically in organizational settings" (p.10). These systems are used by individuals and organisations for making it easier to generate, analyze and use information.

Jessup and Valacich (2003) argue that a computer based information system can be looked as comprising the following important components/elements:

- Hardware: Hardware can be a single PC, a single main frame or networks of computers. It also includes physical device to control the process of input and output like keyboard, mouse and modem.
- Software: Application program such as MS Office, Macromedia Dreamweaver, Koha library software and etc.
- People: Those who are involved with the system or using the system.
- Data: Consists of facts, text, graphic, figure that can be recorded and that have specific meaning

3

- Procedures: Instructions and rules to design and use information system. Strategies, policies, methods, and rules for using the CBIS

Computer based information systems have found application in almost all aspects of our life. Some of the important ones are:
- Science and Engineering
- Business and Commerce
- Education
- Governance
- Health and Medicine
- Entertainment
- Agriculture
- Transport and communication
- Banking and finance

Developing countries have been active in implementing computer based information systems in all these sectors to facilitate service delivery, improve quality of services and cut on costs in the delivery of services. These efforts have not been without challenges. In this paper, I will only focus on the legal challenges to the implementation of computer based systems which have seemingly been overlooked in previous studies.

3.0   LEGAL CHALLENGES IN CONTEMPORARY LITERATURE

3.1 Data Protection and Privacy laws

Computer based information systems are used in such critical areas as electronic funds transfer, internet banking and mobile banking. For these services to be offered effectively, customers' data must be collected in databases and transmitted over long distances electronically. This data must be protected from unauthorised access both during transmission and storage. It is also the responsibility of the party collecting and transmitting this data to ensure that the privacy of the customer in not infringed and that the customers' data is always secure.

The assurance of privacy and integrity is not assured in many developing countries since many developing countries do not have comprehensive data protection and privacy legislation. Therefore, security and privacy concerns by customers sometimes deter them from using computer systems such as ones used to conduct online transactions like buying of goods and services. This is a serious challenge in the implementation of computer systems for these purposes.

Buck (1997) as quoted by Mthembu (2010) is of the same view when he argues that, when making payments, customers want a payment mechanism which is safe and which can enable them to make and receive payments and be sure that no one can divert such payments or impersonate them in order to steal their funds. In the absence of comprehensive data protection and privacy laws that assures them of enough security of their data, their privacy and security of their funds, customers may avoid these systems. This is a serious setback to the development of computer-based systems in the sectors of ebanking, ecommerce and mobile commerce etc.

3.2   Procurement legal framework

Corruption remains a serious obstacle to effective government in Kenya. Public procurement is a major area of corruption: nearly 80% of all cases before the Kenya Anti-Corruption Commission have a procurement element. Corruption in procurement increases the cost of doing business, which in turn adds to the cost of public tenders and leads to poor standards of project work as contractors seek to recover the cost of their bribes.

3.3   Lack of comprehensive anti-corruption legal framework/legislation

Robinson (1998) while making reference to developing countries, states that corruption has become an issue of major political and economic significance in recent years. Corruption is rife in government procurement. Tendering of ICT services and equipment has been a corruption riddled area. For example, Newspapers in Kenya have in many cases pointed out that tenders for ICT services and equipment were being awarded on the basis of personal relations between officials in government and the businesses, instead of on the basis of competence and capability. This results in private sector ICT firms which are not the best in the sector being awarded the tenders. This in turn impacts on the quality of the computer based systems implemented.

3.4   ICT law (e-law) as a field is not yet mature

Like other fields such as property, education, commerce and business, ICT has been identified as a crucial tool for making economies more competitive. Therefore, necessary legislation should be put in place to regulate and accelerate the development of ICT as a major contributor to the economy. May foreign investors shy away from establishing their businesses in developing countries because they feel their interests may not be adequately protected under the current e-law frameworks. For example, existing legislation in many developing countries is not adequate and appropriate for prosecuting software counterfeiting/piracy crimes. This has a serious effect on those developing the software since "pirators" can go unpunished.

3.5   Lack of well-trained lawyers in the field ICT

Lack of lawyers trained in ICT law (E-lawyers) is also a big challenge of implementing computer based information systems in developing countries. For ICT companies planning to set businesses in developing countries, E-lawyers can facilitate this process by advising on matters of compliance with government regulations. This can fasten the process of setting businesses in these countries.

3.6   Tax Laws

Although taxes are essential for sustaining the economy of a country, they can also be a hindrance to the development of some sectors in the economy. In the ICT sector for example, taxes can make products and services unaffordable. Developing countries have tax regimes which make ICT equipment and softwares expensive. For example, re-introduction of VAT on ICT equipment in Kenya has impacted negatively on the growth of the ICT sector and on Kenya's position as the leading digital economy in Sub-Saharan Africa (Omwansa, 2014). This is mainly because the equipment and software prices have skyrocketed and become out of reach to many Kenyans. This is a serious obstacle to adoption of computer based information systems in Kenya especially by small and medium enterprises whose financial resources are limited.

5

The withholding tax imposed on consultancy contracts in Kenya, Uganda and Tanzania for example, means that consultants implementing ICT projects for organisations have to pass on this cost to these organisations which has an effect of making the implementation of these computer based systems more expensive. According to Trademark East Africa, Management or professional fees paid to local consultants is subject to a 10% withholding tax. If a consultant is non-resident, the rate is 20%.

Hassan (2009) in a policy briefing identified excise duty as a contributor to the increase in prices of ICT equipment in Tanzania. He pointed out that, in addition to the standard 20 per cent value added tax, the 10 per cent excise tax on all ICT equipment results into higher consumer prices of the equipment. This hinders the implementation of computer-based information systems due to unaffordability of the ICT equipment.

3.7 Lack of implementation and enforcement of existing laws

Even where proper ICT friendly laws have been passed in developing countries, their benefits have not yet been felt. This is mainly because of lack of effective implementation and enforcement mechanisms. For example, in Kenya, the procurement law was operationalized on 1st January, 2007 vide the Public Procurement & Disposal Regulations, 2006 (Juma, 2010). The new law was aimed at transforming the procurement system by making it more efficient and by eliminating corruption in government's procurement and supply chain. Until now, the law has not been fully implemented. Thus, all sectors that were destined to benefit from these reforms have not yet benefited. Procurement of ICT services and equipment to facilitate the implementation of computer based systems is still taking too long due to bureaucratic procurement procedures. This is a big hindrance to the implementation of computer-based information systems in government sector.

3.8 Lack of National Information & Communications Technology (ICT) Policies

Many Governments countries recognize the role of ICT in the social and economic development of the nation and have promulgated national ICT Policies to fast-track the development of the ICT sector, i.e. the adoption of computer-based information systems. These policies ensure the availability of accessible, efficient, reliable and affordable ICT services. Many developing countries lack these ICT policies and therefore, the development of the ICT sector is ineffective. For example lack of a comprehensive ICT policy may lead to: development (or prolonged existence) of ineffective infrastructure and a waste of resources (Kandiri 2011). To Kandiri (2011), ICT policy, among other things, has the following objectives;

- Improves the quality of services and products
- Provides information and communication facilities, services and management at a reasonable or reduced cost
- Encourages innovations in technology development, use of technology and general work flows
- Identifying priority areas for ICT development
- Etc.

It is therefore apparent that lack of such a policy is a serious impediment to any effort geared towards the implementation and development of computer-based information systems.

3.9   Immigration laws

Availability of qualified and talented manpower in the fields such as software and hardware engineering, database administration, software development and analysis are crucial in the implementation of computer-based information systems. Where such talents are in short supply, it is advisable to recruit beyond borders. This means the immigration laws must be flexible enough in order to allow such recruitment to take place. Some developing countries have strict immigration laws which hinder recruitment of foreign nationals to work locally. This hinders development of computer-based systems in such countries. To take advantage of this emerging trend (recruiting beyond borders), these countries must change their immigration laws changed to open their doors to well-trained talented IT professionals from anywhere. Though this might be seen as competing against the local talents, it may be the only way out in some areas where qualified manpower is next to impossible to get locally.

3.10   Intellectual property laws

The term refers to a range of intangible rights of ownership in an asset such as a software program. Each intellectual property "right" is itself an asset, a slice of the overall ownership pie (Freibrun, n.d).

The law provides different methods for protecting these rights of ownership based on their type.

- **Copyright**

Copyright is "the exclusive legal right, given to an originator or an assignee to print, publish, perform, film, or record literary, artistic, or musical material, and to authorize others to do the same. Implementing computer based information systems especially in libraries, museums, galleries, and archives goes beyond mere installation of hardware, networks and software. It also involves preparing content through digitization and storing information/records in databases and digital repositories and archives.

Copyright law affects what material is digitized, how it is made accessible and how it is accessed to the public. In fact, copyright in some cases prevents libraries from providing open access to the digital information they collect thereby becoming barrier to the development and accessibility of digital collections. Some digital content, for example, will only be accessible to one person at a time due to copyright restrictions. This is in contrast to the spirit of digitization which is to increase access to information resources.

- **Patents**

According to Bridges (2013), Patents are a form of intellectual property. They grant the owner the exclusive right to produce an invention for a specific period. They usually require a 'specification' identifying what the invention does, and how it does it. In many countries, for an invention to be patented,   it must be something new, something that has "an inventive step that is not obvious to someone with knowledge and experience in the subject" and it must "be capable of being made or used in some kind of industry" (Bridges 2013). Patents usually extend to about 20 years. Many industries rely on patents to maintain profitable. Where a patent is granted on a certain type of software application or a computer hardware component, it may create monopoly in as far as that application or component is concerned. This has an effect of making it expensive and unaffordable especially in developing countries. This can also be a serious obstacle to the implementation of computer-based information systems in developing countries.

7

Granting of patents to the creators may also hinder innovation of cheaper and high quality applications and components. This, in itself, is a hindrance to development of computerized systems.

- **Trade Secrets**

Freibrun (n.d) defines a trade secret as any formula, pattern, compound, device, process, tool, or mechanism that is not generally known or discoverable by others, and which is maintained in secrecy by its owner, and gives its owner a competitive advantage because it is kept secret. A good example of a trade secret is the formula to manufacture Coca-Cola. He further argues that many features of software, such as code and the ideas and concepts reflected in it, can be protected as trade secrets. This protection lasts as long as the protected element retains its trade secret status. Trade secrets are powerful economic tools in software industries but they hinder innovation and competition. This has an effect of making the software more expensive. As a result, many developing countries are unable to procure most of the software kept as trade secrets which has an effect of hindering the desire and efforts to implement computer-based information systems.

3.11   Licensing and contract laws

Research in many parts of the world has shown that many content providers who are unhappy with the protections afforded them under copyright law, always turn to contract law and licensing for protection.

Kuny and Cleveland (1998) argue that libraries and other information centres are already at the receiving end. They have huge financial and administrative burdens of managing site licenses for electronic information such as CD-ROMs, ebooks, ejournals and data files. Licensing provides content providers with a stronger mechanism to control the transmission and use of information. This has the effect of moving information from a realm where ideas are allowed to flow in the public domain, to one where this flow is controlled by the provider. This is a serious challenge to digitization and access to digital contents.

4. 0 Conclusion and the way forward

Developing countries face a myriad of challenges in their effort to implement CBIS. These challenges affect all sectors both in government and private settings. The most ignored challenges are usually the legal ones because they are not so apparent to those involved, directly or indirectly, in the implementation of these systems. A recognition of these legal impediments can go a long way in ensuring appropriate legal frameworks are put in place to facilitate smooth implementation of these systems in future.

5.0 REFERENCES

Bridges, C. (2013). *A Case Study: Are patents a help or hindrance to innovation?*. [ONLINE] Available at: http://www.keepcalmtalklaw.co.uk/case-study-patents-help-or-hindrance/. [Last Accessed 28 May 2014].

Freibrun, Eric (n.d). *Intellectual Property Rights in Software: What They Are and How the Law Protects Them.* [ONLINE] Available at: http://www.freibrun.com/articles/articl2.htm. [Last Accessed 4 June 2014].

Hassan, A.K. (2009). *Role of Information and Communication Technologies (ICT) in Enhancing the Livelihood of the Rural Poor.* [ONLINE] Available at: www.taknet.or.tz/topics/ICT_LIVELIHOOD.PDF. [Last Accessed 16 May 2014].

Jessup, L. and Valacich, J., (2004). *Information systems today.* 1st ed. New Delhi: Prentice-Hall of India Private Limited.

Juma,M.J.O. (2010). *Public Procurement Reforms in Kenya.* [ONLINE] Available at: http://www.oecd.org/development/effectiveness/45380758.pdf. [Last Accessed 16 May 2014].

Kandiri, J. (2011). *ICT policy in Kenya and ways of improving the existing ICT policy.* [ONLINE] Available at: http://www.hingx.org/Share/Attachment/894/KENYA'S%20ICT%20POLICY.pdf.. [Last Accessed 4 June 2014].

Kuny, T., & Cleveland, G. (1998). The digital library: myths and challenges. *IFLA journal, 24,* 107-113. [ONLINE] Available at: http://library.nust.ac.zw/gsdl/collect/toolbox/archives/HASH76b7.dir/The%20Digital%20Library%20-%20Myths%20and%20Challenges.pdf. [Last Accessed 16 May 2014].

Mthembu, M. A. (2010). Electronic Funds Transfer: Exploring the Difficulties of Security. *J. Int'l Com. L. & Tech., 5,* 201. [ONLINE] Available athttp://heinonline.org/HOL/LandingPage?handle=hein.journals/jcolate5&div=25&id=&page=. [Last Accessed 17 May 2014].

Omwansa,T. K. (2014). *Re - introduction of VAT on ICT Equipment in Kenya.* [ONLINE] Available at: http://www.c4dlab.ac.ke/wp-content/uploads/2014/04/VAT_Report_TKO-Executive_Summary.pdf. [Last Accessed 16 May 2014].

Robinson, M. (1998). Corruption and development: An introduction. *The European Journal of Development Research, 10*(1), 1-14.

TradeMark East Africa (2012). *Guide on Withholding Tax (WHT) for Consultancy Contracts.* [ONLINE] Available at: http://www.trademarkea.com/wp-content/uploads/2012/09/Guide-on-Withholding-Tax-EA-for-Consultancy-Contracts.pdf. [Last Accessed 16 May 2014].